TH

APR 2 3 1985

HOW DID WE FIND OUT ABOUT
THE BEGINNING
OF LIFE?

HOW DID WE FIND OUT ABOUT

THE BEGINNING OF LIFE?

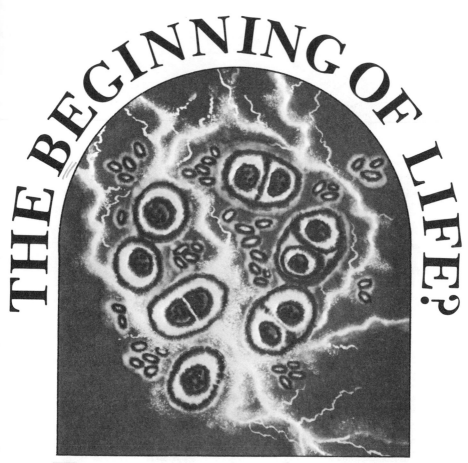

Isaac Asimov
Illustrated by David Wool

WALKER AND COMPANY
New York

Library of Congress Cataloging in Publication Data

Asimov, Isaac, 1920–
  How did we find out about the beginning of life?

  (How did we find out—series)
  Includes index.
  Summary: Describes scientists' attempts to find out how life be-
gan, including such topics as spontaneous generation and evolution.
  1. Life—Origin—Juvenile literature. [1. Life—
Origin] I. Wool, David, ill. II. Title.
III. Title: About the beginning of life.
QH325.A74   1982   577   81-71196
ISBN O-8027-6447-9   AACR2
ISBN O-8027-6448-7 (lib. bdg.)

First published in the United States of America
in 1982 by the Walker Publishing Company, Inc.

Published simultaneously in Canada by John Wiley & Sons Canada, Limited,
Rexdale, Ontario

Trade ISBN: 0-8027-6447-9

Reinf. ISBN: 0-8027-6448-7

Library of Congress Catalog Card Number: 81-71196

Printed in the United States of America

10 9 8 7 6 5 4 3 2 1

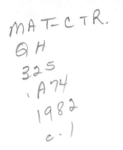
To
Hazel Jeppson Price,
with a nephew-in-law's affection.

HOW DID WE FIND OUT . . . ? SERIES
Each of the books in this series on the history of
science emphasizes the process of discovery.

Contents

1 Spontaneous Generation

WE ALL KNOW that people have babies, dogs have puppies, and cats have kittens. When we visit the zoos, we find out that bears have cubs and deer have fawns and so on.

Every baby animal comes from a living animal mother, who was born from another, still earlier animal, and so on.

You yourself have a mother, and your mother was once your grandmother's baby, and your grandmother was once your great-grandmother's baby, and so on.

Some animals, such as birds, lay eggs. Every bird alive came out of an egg, which had been laid by a parent bird, which once came out of an egg, which had been laid by another parent bird, and so on.

It's the same with plants. If you want to grow plants, you must plant seeds that were produced by plants that had grown previously. And those previous plants were grown from seeds produced by previous plants, and so on.

1

Where did it all start? Does it go back forever? Or was there a time when there was an original human being and dog and cat and bear and chicken and daisy?

If that is the case, how did the original living thing come into being?

Before modern times people didn't really think that was much of a mystery. At least, they didn't think it was a mystery in the case of *some* living things.

Some forms of life just grew, or appeared, seemingly from nowhere. This was usually the case with living things that annoyed us or that were of no use to us.

For instance, very few people are interested in crocodiles and snakes, and very few people want them. In fact, most people try to kill them. Yet they keep appearing.

In William Shakespeare's play, *Antony and Cleopatra,* one of the characters is Lepidus, a Roman general. Shakespeare has him say: "Your serpent of Egypt is bred now of your mud by the operation of your sun; so is your crocodile."

Some people may have believed that crocodiles and snakes were formed out of mud that had been heated by the sun, but, of course, that is not so. Crocodiles and snakes lay eggs, and these hatch into baby crocodiles and snakes.

But what about smaller and still more numerous creatures?

In the days before refrigeration, it often happened that meat spoiled and grew rotten. In that case, tiny wormlike *maggots* would appear on the meat.

2

It looked as though live maggots formed from dead meat. Life seemed to form from nonlife all by itself. If maggots could do it, surely other forms of life could do it, too, under the right circumstances. Perhaps thousands of years ago, even snakes, crocodiles, chickens, dogs, and human beings were formed from nonlife.

This formation of life from nonlife is called *spontaneous generation* (spon-TAY-nee-us jen-uh-RAY-shun). This means "the forming of life without outside help."

In old times scholars all took spontaneous generation for granted.

In 1668, however, an Italian physician, Francesco Redi (RAY-dee, 1626–1697), thought that the notion ought to be tested. After all, what if small things that were alive laid eggs on spoiled meat? The eggs might be so small that people couldn't see them, and out of these invisible eggs, maggots might come.

Redi therefore put fresh meat in eight different flasks. He sealed four flasks tightly so that nothing could get at them. He left the other four flasks open so that flies, for instance, could buzz about the meat and settle on it.

As the days passed, the meat in the open flasks grew rotten and smelly, and maggots began to crawl over it. When Redi opened the sealed flasks, the meat was just as rotten and smelly, but there were no maggots.

Could it be the absence of fresh air that kept the maggots from forming? Redi tried another experiment. He put fresh meat in flasks that he left open but

3

FRANCESCO REDI

MAGGOT EXPERIMENT

with the opening covered by gauze. Fresh air could get in, but not flies. The result? The meat turned rotten, but no maggots formed.

The conclusion was clear. Flies laid eggs, and maggots hatched out of the eggs, and eventually became flies themselves, just as caterpillars became butterflies.

That was a big point against spontaneous generation.

At the time that Redi made this discovery, scientists were just beginning to use *microscopes,* which enlarged tiny things and made them visible.*

A Dutch scientist, Anton van Leeuwenhoek (LAY-ven-hook, 1632–1723), using a microscope, discovered living things in 1675 that were too small to see without one. These are now called *microorganisms* (MY-kroh-AWR-guh-niz-umz). He watched them move about and eat other microorganisms.

Where did these microorganisms come from? Most were less than a hundredth of an inch across. Could they lay eggs?

*See *How Did We Find Out About Germs?* (New York: Walker, 1974).

5

LIFE CYCLE OF FLY

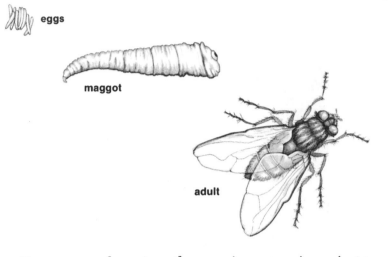

eggs

maggot

adult

One way of seeing these microorganisms is to get some water out of a ditch or pond. If a broth, made by soaking food in water, is added to the ditchwater, the microorganisms feed on the broth. They grow and multiply.

However, nothing has to be added to the broth at all. The broth might be freshly prepared and filtered, and, if it is studied under the microscope, it might be seen to have no microorganisms in it. Yet if it is allowed to stand awhile, it will be found to be swarming with microorganisms.

Surely that is an example of spontaneous generation. The living microorganisms formed out of the dead broth. Or did they?

It could be that there are microorganisms floating around in the air. If some happened to fall into the broth, they would multiply there.

To check this notion in 1748 an English scientist, John T. Needham (NEED-um, 1713-1781), began with fresh mutton broth. He boiled it in a flask in order to kill any microorganisms that were already there. Then, while the broth was still hot, he sealed the flask. A few days later, he opened the flask and studied the broth under a microscope and found it full of microorganisms. He announced that this proved spontaneous generation had taken place, since nothing could have fallen in after the flask had been sealed.

One person who was not convinced by this was an Italian scientist, Lazzaro Spallanzani (spahl-lahn-TSAH-nee, 1729-1799). He wondered if Needham had really killed all the microorganisms to begin with. After all, he had only boiled the broth a few minutes.

Spallanzani tried the experiment again, in 1768, but *he* boiled his broth for over half an hour. Then he sealed the flasks. It turned out that no matter how long he left them sealed, no microorganisms were found in them when they were opened. Spallanzani insisted that there were microorganisms floating in air and that these were the source of any tiny living things that appeared in broth.

Spallanzani studied individual microorganisms under the microscope and observed one dividing into two living microorganisms. There were no eggs. The microorganisms just divided in two. That is how they multiply.

But are there really microorganisms floating in air at all times? A German scientist, Theodor Schwann

THEODOR SCHWANN

(SHVAHN, 1810–1882), tested that notion in 1836. He boiled broth just as Spallanzani had done, but he didn't seal the flask. Instead he exposed it to a current of air, but he heated that current of air strongly enough to kill any microorganisms that might be in it.

No microorganisms appeared in the broth.

Some scientists thought there might be something in the air—a vital principle—that made it possible for spontaneous generation to take place. It might be something that would be destroyed by strong heat, and then the dead broth could no longer give rise to living microorganisms.

To check that, a French chemist, Louis Pasteur (pas-TER, 1822–1895), tried a new experiment in 1860.

He boiled broth until everything in it was killed, but he kept the broth in a flask with a long, thin neck. The neck went up in the air, then bent to one side and down, and then up again, like the letter S lying on its side.

Once the broth cooled down, cool air could drift inward through the long, thin neck, and that air ought to be full of the vital principle, if there were any such thing.

Only air came in. Any dust in the air settled out in the lower part of the downward bend of the neck. Pasteur felt that any microorganisms in the air would be attached to the dust particles and would also settle out there. They did, and the broth developed *no* microorganisms. If Pasteur broke off the neck, however, so that air plus dust could reach the broth, microorganisms began appearing at once.

LOUIS PASTEUR

After Pasteur's experiment, the notion of spontaneous generation seemed to be dead. A German scientist, Rudolf Virchow (FIHR-khuv, 1821–1902), when he heard of the experiment, said, "All life comes from life." That seemed to be the rule that scientists accepted thereafter.

2 Evolution

NOT ONLY DOES all life come from life, but all life seems to come from the *same* life. Dogs always have puppies, and cats always have kittens. Beavers always produce young beavers. Ostriches always lay eggs, out of which come baby ostriches. Oak trees always produce acorns, out of which grow more oak trees, and so on.

Every kind of plant or animal or microorganism that produces only plants, animals, or microorganisms of the same kind as itself is called a *species* (SPEE-sheez).

There is only 1 species of human beings, but there are 2 species of elephants, the Indian elephant and the African elephant. There are 3 species of hyenas, 8 species of badgers, 9 species of foxes, 500 species of fleas, and at least 660,000 species of other insects.

Scientists have discovered about a million different species altogether, and there may be another million (mostly insects and other small beings) that have not yet been discovered.

Since this is so, it would seem that scientists have to do more than puzzle out how life got started. They have to figure out how each of two million different kinds of life got started.

Did they all get started at the same time? In the same place? In the same way? Or were conditions different for each one?

As it happens, though, various species aren't all equally different. Some of them form groups of similar species and then groups of similar groups, and so on.

For instance, there are different species of wolves and of foxes, but they are all doglike animals. Lions, tigers, leopards, and jaguars are all catlike animals. These doglike animals and catlike animals, along with bears, weasels, seals, and so on, are all meat-eating animals, or *carnivores.*

In contrast to the carnivores are plant-eating animals, or *herbivores,* such as sheep, deer, rabbits, mice, and so on. But they resemble meat-eating animals in having hair and warm blood and in producing live young that feed on milk. All of these animals—both carnivores and herbivores—are *mammals.*

Then there are many species of birds and of reptiles and of fish. They are not mammals, but they resemble mammals in having bones. They and the mammals are all *vertebrates.*

Before modern times not much was done in the way of classification, but, beginning in 1660, an English naturalist, John Ray (1628–1705), studied and classified about 18,600 different species of plants. He

JOHN RAY

divided them, to begin with, into two groups. One group included those plants with seeds containing one tiny seed leaf. The other group included those plants with seeds containing two seed leaves.

In 1693 he also classified animals. He divided them, for instance, into those that had hooves and those that did not. He divided those with hooves according to whether they had one, two, or three hooves on each leg.

Even more important was the work of a Swedish naturalist, Carolus Linnaeus (lih-NEE-us, 1707–1778). In 1735 he published a book in which he classified plants and animals very neatly. He grouped similar species into *genera* (JEN-uh-ruh), similar genera, into *families*, similar families into *orders*, and similar orders into *classes*.

In later years, a French scientist, George Cuvier (koo-VYAY, 1769–1832), grouped similar classes into *phyla* (FY-luh) and similar phyla into *kingdoms.*

This sort of classification worked well. What's more, it seemed to arrange all living things into something that resembled a tree.

The trunk of the tree is life itself. The trunk divides into four kingdoms: animals, plants, and two different kinds of microorganisms. Each kingdom divides into several phyla, each of which divides into several classes, then orders, families, and genera. Finally, the genera divide into the separate twigs that make up the two million living species.

When scientists considered the *tree of life,* some of them couldn't help wondering if perhaps the whole arrangement grew the way a real tree did. Was there

CAROLUS LINNAEUS

once an original vertebrate, for instance, that gave rise to mammals, birds, reptiles, and so on? Was there once an original mammal that gave rise to all the different mammals that now exist? Did one species slowly change into another species or into a whole group of similar species?

This notion of one species changing into another is called *evolution* (eh-voh-LOO-shun).

Of course, no one sees any species changing. All through history, cats have remained cats, and dogs have remained dogs. History is only about five thousand years old, however. Perhaps such changes are very slow and take much, much longer than five thousand years.

As the 1800s proceeded, scientists became convinced that planet Earth was many millions, even hundreds of millions, of years old and that there was plenty of time for evolution to take place, even if it proceeded very slowly. In fact, nowadays scientists think that Earth is about 4.6 billion (4,600,000,000) years old.

But then, *why* should species change? Even if we suppose it happens very slowly and that there is lots of time for it to happen, *why* should it happen?

The first person to suggest a reason for it was a French naturalist, Jean de Lamarck (lah-MARK, 1744–1829). In 1809 he published a book in which he suggested that species changed because each plant or animal changed in the course of its life, and its young inherited the changes.

For instance, some short-necked antelope ate leaves and kept stretching its neck to reach leaves

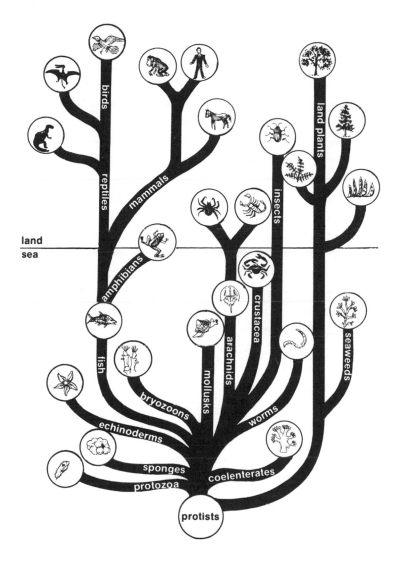

THE TREE OF LIFE

19

higher in the tree. Its neck stretched permanently in this way, just a little bit. Its offspring inherited that trait, having a neck that was a tiny bit longer than that of its parents. This kept on, generation after generation, for thousands of years, and finally the antelope became a giraffe. In the same way, some species become faster, or larger, or smaller, and so on.

However, organisms do not pass on to their young any changes they acquire. This was tested over and over again, and Lamarck was found to be wrong.

A much more useful suggestion was made by an English naturalist, Charles R. Darwin (1809–1882). In 1859 he published a book called *The Origin of Species*, in which he pointed out that different members of a particular species always differed slightly among themselves. Some might be stronger than others, or faster, or darker in coloring, or have sharper eyes, a better nose, and so on.

Those animals that could catch food more easily, or fight off enemies more successfully, or hide from enemies more skillfully, or endure starvation better would live longer and have more young. They would pass on their characteristics to their young because they would not have acquired those characteristics during their lifetime but would have been *born* with them.

This would happen generation after generation and, in this way, species would slowly change to fit the environment. Different species would evolve that would adapt in different ways. One would run better, or hide better, or fight better.

JEAN DE LAMARCK

Darwin's suggestion of *evolution by natural selection* succeeded. More and more scientists found more and more evidence in its favor. Since Darwin's time his notions have been greatly improved upon, and the fine points of evolution are still being argued about even now. However, scientists today are quite sure that species have evolved from other species.

CHARLES DARWIN

3 The Earliest Life

SCIENTISTS not only accept the notion that species have evolved from other species. They even believe they know a great many of the details of the process.

Over the long period of time that life has existed, there have been times when animals have died and been covered by mud before they could be eaten. The mud hardened in time, and the bones or shell or skin of the animal (or the wood of plants) have slowly changed into rock. Some of these rocky formations can be dug out and are then found to have just the shape of the original animal or plant parts that were buried. These rocky formations are called *fossils* (FOS-ilz).*

Some of these fossils are tens of millions or even hundreds of millions of years old. They are of different species from those that are alive now. However, the fossil species that are now "extinct" (that is, no longer alive) can be fitted into the same arrangement as modern species can.

*See *How Did We Find Out About Dinosaurs?* (New York: Walker, 1973).

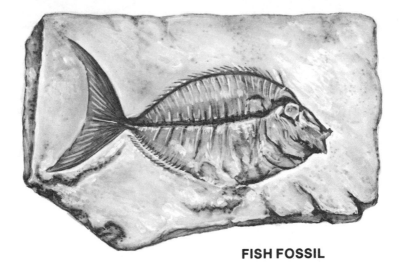

FISH FOSSIL

There are a whole series of fossils of horselike animals, for instance. If these are arranged in order of age, we begin with a small animal with four hooves (toes) on each of its front legs. As time went on, the animal changed from one species to another, growing larger, having longer legs and fewer hooves, until the large, horselike animals of today, with one hoof on each leg, developed.

There are also fossils of gigantic animals that lived a hundred million years ago. They are reptiles, just as modern crocodiles and lizards are reptiles, but they are much larger. These long-extinct giant reptiles are the animals that we usually speak of as *dinosaurs*.

There are fossils of an animal that had the tail and teeth of a lizard, but feathers like a bird. It seems to be a species that descended from reptiles and was the ancestor of birds.

Scientists learned how to judge the age of fossils more and more accurately, and the oldest fossils of plants and animals large enough to be seen without a microscope are about six hundred million (600,000,000) years old.

Back then there were no human beings. There were no cats or dogs or birds or snakes or fish. There were no animals with bones at all. In fact, there were no animals that lived on land.

The only animals that existed were those that lived in the sea, and the most complicated ones were called *trilobites* (TRY-luh-bites).

So you see, if we're going to wonder about how life started, we don't have to wonder about each of the two million species now alive. We can wonder about the fewer and simpler species that lived hundreds of millions of years ago.

Yet that's not good enough.

Even six hundred million years ago there were still quite a number of different species, and the trilobites were pretty complicated animals. They were much more complicated than some of the smaller and simpler species of today.

We have to wonder how the trilobites got started.

The oldest fossils are six hundred million years old but the Earth is over seven times as old as the oldest fossils, and life could have existed for much longer than the fossils indicate. If there was life long before the trilobites, however, why didn't it leave fossils behind?

Actually, fossils are mostly of the parts of plants and animals that turn to rock easily. Fossils are

EVOLUTION OF THE HORSE

Eohippus	**Mesohippus**	**Merychippus**
58 million years ago	**36 million years ago**	**25 million years ago**

usually formed from what were once bones and teeth and shells and wood—the hard parts of living things.

These hard parts seem to have evolved very late. When the trilobites first existed, no animal had yet developed bones, for instance, and no plants had developed much in the way of wood.

Longer than six hundred million years ago, shells hadn't yet been formed either. No hard parts were formed. Plants and animals were small and soft and didn't leave fossils. In fact, to begin with, the only living things on Earth must have been microorganisms, tiny bits of life only a hundredth of an inch across or less.

Such microorganisms are formed of only a single bit of life called a *cell*. It was only later on, perhaps much later on, that cells grouped together in the course of evolution to form larger *multicellular* (MUL-tee-SEL-yoo-ler) *organisms*.

Pliohippus
13 million years ago

Equus
1 million years ago

With the passing of time, organisms formed that contained millions and billions of cells, and even more. (A human being contains fifty trillion—50,000,000,000,000—cells.)

With more and more cells, it is possible for groups of them to specialize into different organs—into eyes and muscles and stomachs and shells and bones.

The first forms of life had none of this, however. They were just tiny single cells and could leave no fossils of the ordinary kind.

Just the same, in very old rocks, scientists have found microscopic markings that look as though they might be all that is left of very ancient cells.

In 1965 an American scientist, Elso S. Barghoorn (BAHRG-hawrn, 1915–   ), found such *microfossils* in rocks that were over three billion years old.

Nowadays, scientists think that life began on Earth perhaps as long as three and a half billion

27

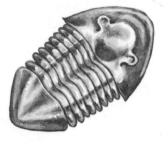

TRILOBITE

(3,500,000,000) years ago, or when the Earth was only about a billion years old. Life has been developing and evolving ever since.

When we ask how life started, then, we are not asking how trilobites started. We are asking: How did those tiny microscopic bits of life start over three and a half billion years ago?

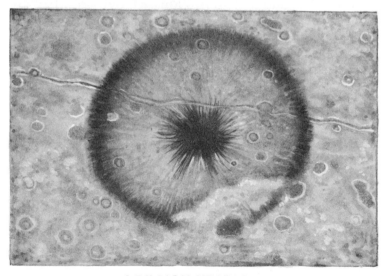

**2 BILLION YEAR OLD
FOSSIL ALGAE IN CANADIAN ROCK**

4 Proteins and Nucleic Acids

IF ONE SPECIES developed from another, and if all species evolved from some very simple life form that existed three and a half billion years ago, then all the many millions of species, alive and extinct, must resemble each other somehow.

They do. All living things (and all nonliving things, too) are made up of tiny atoms.* These atoms are grouped together into *molecules*, and the molecules in all living things are remarkably similar to each other. The molecules in tiny microorganisms are very much like those in rats and lobsters and oak trees and herrings and rose bushes and human beings. There are differences in details, of course, but the general likenesses are the strongest arguments in favor of evolution.

In the late 1700s, chemists began to study the molecules in living things. The English chemist William Prout (1785–1850) in 1827 divided them into three chief classes. In the first class were starches and

---

*See *How Did We Find Out About Atoms?* (New York: Walker, 1976).

sugars; in the second class were fats and oils; and in the third class were certain substances such as those in egg white. This third class was first referred to as *albumins* (al-BYOO-minz) from the Latin word for "egg white."

The molecules of starches, sugars, fats, and oils are all made up of atoms of carbon (KAHR-bon), hydrogen (HY-druh-jen), and oxygen (OK-sih-jen). The molecules of albumins also contain those atoms, but in addition they contain atoms of nitrogen (NY-truh-jen) and sulfur (SUL-fur).

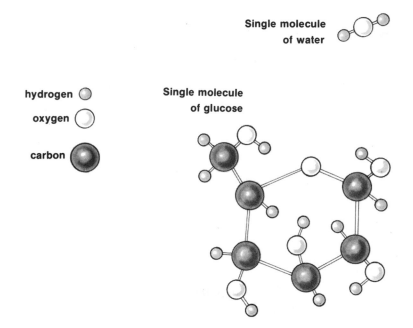

Single molecule of water

hydrogen

oxygen

carbon

Single molecule of glucose

MOLECULES OF WATER AND GLUCOSE

The albumins seemed much more complicated than the other compounds and, in 1838, a Dutch chemist, Gerardus J. Mulder (1802–1880), called them *proteins* (PROH-tee-inz). This is from a Latin word meaning "first," showing that they seemed to be of first importance in living things.

As time went on, proteins proved to be complicated indeed. The molecules of many of them were made up of tens of thousands, even hundreds of thousands, of atoms.

The atoms in protein molecules aren't put together in any old way. The protein molecules are long chains of simpler molecules called *amino acids* (uh-MEE-noh-AS-idz).

An amino acid molecule of the type usually found in proteins is made up of from ten to twenty-two atoms. All of them contain carbon, hydrogen, oxygen, and nitrogen atoms. Some contain sulfur atoms in addition.

There are twenty different amino acids that each appear in almost every protein molecule. They can be arranged in any order in making up the protein chain, and every different order results in a protein molecule that is slightly different in its properties from one with any other order. That means that there are an enormous number of different protein molecules possible.

Suppose you had four different amino acids and numbered them 1, 2, 3, and 4. You could arrange them as 1-2-3-4, or 1-2-4-3, or 2-3-4-1 or 3-4-2-1. There would actually be twenty-four different arrangements possible.

AMINO ACIDS LINKED END TO END

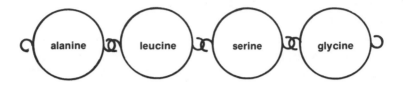

If you started with twenty different amino acids, you could arrange them in more than twenty-four billion billion (24,000,000,000,000,000,000) ways. Actual protein molecules can be made up of dozens of each of the twenty amino acids, so that the number of different proteins possible is far, far greater than the number of all the atoms in the universe.

It is the tiny differences in amino acid order that make it possible for all living things to contain protein molecules and yet for them to be as different as a daisy is from a whale. Daisies and whales both have proteins built up of amino acids—but in different orders.

What keeps the amino acid order just so? Why should a daisy seed always produce a living thing with daisy proteins in it? Why should a whale always produce a living thing with whale proteins in it?

It was a long time before any answer was found for these questions.

The beginning of the answer came in 1869, when the Swiss chemist Johann F. Miescher (MEE-sher, 1844–1895) found a new substance in a little structure usually present at the center of a cell. This structure is known as the cell's *nucleus* (NOO-klee-us), so the substance Miescher discovered came to be known as *nucleic* (noo-KLEE-ik) *acid.* Nucleic acid molecules contain not only carbon, hydrogen, oxygen, and nitrogen atoms, but atoms of phosphorus (FOS-fuh-rus) in addition.

SINGLE CELL

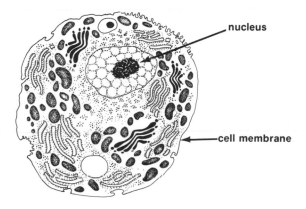

Nucleic acids, like proteins, have molecules that are built up of chains of small molecules. The nature of the small molecules was not known until 1909, when a Russian American chemist, Phoebus A. T. Levine (1869–1940), worked it out. These small molecules are called *nucleotides* (NOO-klee-oh-tidez) and have about forty atoms apiece.

In any nucleic acid there are only four different nucleotides, but it turned out that nucleic acid chains are so long that even with only four, the total number of different arrangements is every bit as great as in the case of proteins.

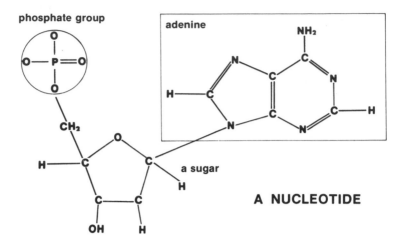

phosphate group

adenine

a sugar

A NUCLEOTIDE

In 1944 a Canadian scientist, Oswald T. Avery (1877–1955), was able to show that nucleic acids are even more important than proteins. He could change one kind of microorganism into another very similar kind by putting a kind of nucleic acid called DNA from the second into the first. Proteins wouldn't do the trick. Until that time most scientists had ignored nucleic acids and thought them not very important. Now, however, they began to study them thoroughly.

In 1953, an English scientist, Francis H. C. Crick (1916–   ), and an American one, James D. Watson (1928–   ), working together, showed just how a nucleic acid was shaped. They showed how any nucleic acid molecule could form another exactly like itself.

Since the nucleic acid molecules control the shape of the protein molecules, and the protein molecules control the nature of living things, you can see what

34

FRANCIS H. C. CRICK   **JAMES D. WATSON**

must happen. Nucleic acids in a living organism produce others exactly like themselves, and some of these are passed on to the young. The nucleic acids in the young then produce proteins just like those in the parents, so that the young are just like their parents.

It is because nucleic acids produce themselves so exactly that dogs have puppies and cats have kittens, and never vice versa.

Sometimes, though, nucleic acids don't duplicate themselves completely accurately. A wrong nucleotide may get into place here or there, and this can produce a very small difference, or *mutation*. The difference is so small that a puppy is still very much a puppy, but it can have some tiny difference that marks him off from others in the litter. It is because of these tiny mutations that are taking place all the time that every one of the billions of human beings has his or her own face, voice, and appearance so that we can tell them apart.

It is these mutations that make evolution possible by giving natural selection something to act on.

As far as scientists have been able to tell, every one of the species of life they have studied, from the largest to the smallest, contains proteins and nucleic acids.

We can assume, then, that the very first forms of life, three and a half billion years ago, were made up of proteins and nucleic acids.

If we ask the question again, then, about how life began, we are really asking: How did the first proteins and nucleic acids come to be formed, and how did they form the first living thing?

5 The Early Atmosphere

WAIT A BIT. If we're asking how proteins and nucleic acids first came to be formed and how they became living things, aren't we talking about spontaneous generation? And didn't Pasteur show that spontaneous generation was impossible?

Well, Pasteur didn't quite show it was impossible after all.

He showed that spontaneous generation didn't take place in his flask over a period of some weeks, or possibly years, if he waited long enough. However, no life may have appeared on Earth until as much as a billion years had passed. Perhaps, if we could wait a billion years, we would find that life had formed in Pasteur's flask, too.

Well, then, if we study various places on Earth, places that have been left alone for a billion years, might not we find life forming out of nonlife today?

No! The Earth today is full of living things almost everywhere—in the water and on the land, in the ocean surface and in the ocean deeps, on mountains and in valleys, even in deserts.

If protein molecules or nucleic acid molecules were to appear today, some form of life would surely eat them at once, and that would be the end of them. They'd be gone long before they could develop to the point of a living organism.

Three and a half billion years ago, however, there was no life on Earth. If proteins or nucleic acids formed in the early ocean, they would just remain there. There would be nothing to eat them. The ocean would accumulate more and more of them. They might grow more and more complicated and, finally, life would start.

Once nucleic acids and proteins got complicated enough to be living and combined with each other to form primitive cells, they would begin to eat the chemicals about them and to multiply. The cells would vary among themselves, and natural selection would see to it that some would flourish and some would die out. Evolution would begin, and that would be the start of the long process that would produce the world of today—and us.

But how could the proteins and nucleic acids start in the first place? If they started to form all by themselves from simpler, nonliving molecules, the oxygen in the air about us would probably destroy them as fast as they appeared.

The oxygen in the atmosphere wasn't always there, however. The oxygen in the atmosphere was formed by plants, which are absorbing carbon dioxide from the air all the time and giving off oxygen.

Right now, because of the action of plants, Earth's atmosphere is four-fifths nitrogen and one-fifth oxy-

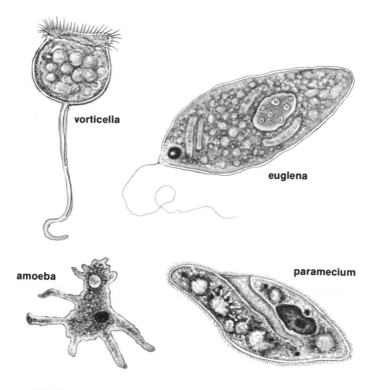

SINGLE-CELLED ANIMALS (MICROORGANISMS)

gen. Only one-three-thousandth of the atmosphere is carbon dioxide (which has a molecule made up of one carbon atom and two oxygen atoms). A billion years or more ago, before there were any plants, there was no oxygen in the atmosphere. In its place there was carbon dioxide. Earth's atmosphere was a mixture, then, of nitrogen and carbon dioxide. The atmosphere on the planets Mars and Venus, where there is no life, is a mixture of nitrogen and carbon dioxide today.

39

That might not have been the original atmosphere of Earth, however. The sun and the giant planets, such as Jupiter and Saturn, consist mostly of hydrogen. The cloud of dust and gas out of which the entire solar system formed was probably mostly hydrogen, plus the combination of hydrogen atoms with other kinds of atoms.

The most common combinations would have been methane (METH-ane, four hydrogen atoms and a carbon atom), ammonia (uh-MOH-nee-uh, three hydrogen atoms and a nitrogen atom), water (two hydrogen atoms and an oxygen atom), and hydrogen sulfide (two hydrogen atoms and a sulfur atom).

When Earth was first formed, it couldn't hold on to the very small and light hydrogen molecules (two hydrogen atoms apiece, and nothing more), but it would have held on to the others. The original ocean would be water in which a great deal of ammonia and hydrogen sulfide was dissolved, while the air would be mostly methane, with some ammonia, hydrogen sulfide, and water vapor also present.

Sunlight shining on this atmosphere would slowly break up water molecules into hydrogen and oxygen, and the oxygen would combine with the methane and ammonia, slowly converting them into carbon dioxide and nitrogen. Then, after plants had developed, the carbon dioxide would be changed to oxygen.

In this way, it may be that Earth has had three different atmospheres. We are living in the third atmosphere (nitrogen and oxygen), but life may have begun in the second atmosphere (nitrogen and carbon dioxide), or even in the first (ammonia and methane and hydrogen sulfide).

The first to suggest that life might have begun in an atmosphere different from the one we now have was an English chemist, John B. S. Haldane (1892–1964). He made the suggestion in 1929.

Then, in 1936, a Russian chemist, Alexander I. Oparin (1894–), went into the matter in greater detail. He thought life might have begun in the first atmosphere.

Methane, ammonia, water, and hydrogen sulfide are all small molecules, with three to five atoms apiece. Among them they have carbon, hydrogen, oxygen, nitrogen, and sulfur atoms, which can combine into all the amino acids, which are larger molecules.

There is a catch. Generally, small molecules are more stable and tend to avoid breaking up more than larger molecules do. For that reason, small molecules don't usually combine, all by themselves, to form larger molecules. Quite the reverse! Large molecules tend to break up into smaller pieces.

Going from large molecules to small ones would be like rolling downhill. To expect small molecules to form large ones all by themselves would be like having them roll uphill. Small molecules would have to be driven to do it, and what can drive them uphill to form larger molecules, and, eventually, life?

Energy can do it.* On Earth, in its very early days, there was a good supply of energy. There were lightning bolts, there was volcanic heat, and there was, of course, sunlight. Nowadays, ultraviolet light,

EARLY STAGE IN THE HISTORY OF THE EARTH

which contains more energy than ordinary light, doesn't reach the surface of the Earth much. There is a layer of ozone (OH-zone, a form of oxygen) fifteen miles high in the atmosphere, and it stops the ultraviolet. In the days when life first began, however, there was no oxygen in the atmosphere, and therefore no ozone. The ultraviolet light reached the Earth's surface in full strength.

Thanks to energy, the small molecules might be able to move uphill to form large molecules, and eventually life might begin.

*See *How Did We Find Out About Energy?* (New York: Walker, 1975).

6 Experiment

It isn't enough to think that perhaps the Earth's atmosphere was of this kind or that, and perhaps energy could do thus and so, and perhaps life could form. Is there any way in which we could check on the matter?

Well, we can't get a time machine and go three and a half billion years back in time, but perhaps there are other ways of doing it.

One person who was particularly interested in the chemistry of the early Earth and in the possible origin of life was the American chemist Harold C. Urey (YOO-ree, 1893–1981). He wondered if, perhaps, the conditions that existed on the early Earth could be imitated in the laboratory today. It might be possible, in that case, to watch what would happen.

Urey had a student, Stanley L. Miller (1930–   ). In 1952 Urey asked him to try the experiment.

Miller began with pure water, which he heated to make sure there was no life of any kind in it. He added hydrogen, ammonia, and methane to it, and

43

STANLEY L. MILLER

in that way he set up a gas mixture that might be like that of the early atmosphere.

He kept this mixture of water and gases circulating through his apparatus, and at one point a discharge of electricity would pass through it. This would be a source of energy similar to that of lightning.

He kept this up for a week and, by the end of that time, the water had turned pink, so there must have been some change in it. At the end of the week, he opened his apparatus and carefully analyzed the contents.

There were no living things in it, of course, but there *were* molecules present that were more complicated than the ones he had begun with. One sixth of the methane had formed more complicated mole-

MILLER'S APPARATUS

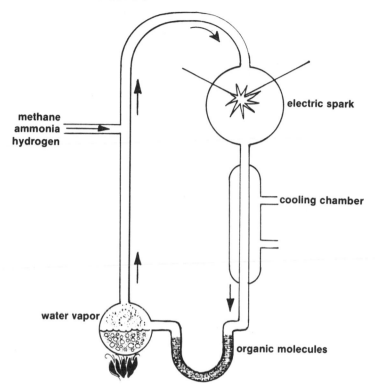

methane
ammonia
hydrogen

electric spark

cooling chamber

water vapor

organic molecules

cules. The energy of the electric discharge had driven the methane uphill. In fact, small quantities of two of the simpler amino acids found in proteins were present.

If two amino acids could be formed in a small flask of water in just a week, what could we expect to happen in a whole ocean of water in a billion years?

Other chemists followed Miller. The American chemist Philip H. Abelson (1913–) tried a variety of different mixtures of simple compounds. He found that no matter what the mixture, as long as there were carbon, hydrogen, oxygen, and nitrogen atoms present, he would end up with amino acids.

In 1959 two German chemists, Wilhelm Groth and H. von Weysenhoff, used ultraviolet light instead of electric discharges as a source of energy—and amino acids were still formed.

Suppose chemists started with larger quantities and continued the experiments for longer times? Would they get still more complicated atoms? They did.

Or suppose they took some of the compounds that formed, and included them among the starting materials. In 1961 the Spanish American chemist Juan (hwan) Oro added hydrogen cyanide (one atom of hydrogen, one of carbon, and one of nitrogen) to the starting mixture. After all, hydrogen cyanide had been formed in Miller's original experiment.

As a result, more amino acids were obtained. In fact, some of the amino acids hooked together to form very short chains. Oro also formed *purines* (PYOO-reenz), a kind of molecule that makes up part of the nucleotides that form nucleic acids. In 1962 Oro added formaldehyde (one carbon atom, two hydrogen atoms, one oxygen atom) to the starting mixture, and he got sugar molecules that are also parts of nucleotides.

In 1963 the Ceylonese American chemist Cyril Ponnamperuma (pon-AM-puh-ROO-ma, 1923–) began with a number of substances that had been previously formed in such experiments, together with a simple phosphorus-containing compound. He succeeded in forming whole nucleotides, and even two nucleotides hooked together.

The American chemist Sidney W. Fox (1912–) went at it in a different way. In 1958 he started with amino acids and subjected them to heat in the absence of water. The amino acids hooked together to form proteinlike molecules. When these were dissolved in hot water, they clung together in tiny spheres like little cells.

All the experiments that were carried on since Miller's first one seemed to show that the changes that took place were always in the direction of life. The chemicals that were formed always resembled those in living things.

It seemed that the appearance of life on Earth was no miracle at all. It was not even surprising. Given the starting chemicals and a source of energy, things would just naturally move in the direction of life.

In that case, we might argue that life would exist on any planet on which it had the slightest chance of appearing. If this is so, we might be able to find life on some other world.

Unfortunately, the worlds we can reach are so different from Earth that life doesn't have a reasonable chance. The Moon has no air or water; Mercury and Venus are almost red hot. The worlds beyond Mars are extremely cold, and their chemistry is altogether different from Earth's.

Mars seemed the best bet. Its air was very thin; it had very little water, and it was very cold. Still, perhaps simple life forms existed. Or, if not that, then perhaps there were chemicals in the soil that were partly on the way to life—amino acids, for instance.

In 1976 two rocket-powered probes reached Mars, landed on its surface, and tested the soil. They could find no traces of molecules containing carbon atoms, and without such molecules there can't be any life like that on Earth.

However, there are some bits of otherworldly matter that actually come to Earth—meteorites that fall to the planet from outer space.

Most meteorites are metallic or rocky and do not have the same elements that living things have. Once in a while, though, a rare type of meteorite containing small amounts of water and carbon compounds arrives.

VIKING ON MARS

In 1969 such a meteorite fell in Australia, and many pounds of fragments were quickly collected. These were carefully studied by chemists, including Ponnamperuma. They found that the organic matter in the meteorites contained eighteen different amino acids, of which six were among those that occur in proteins in living things. This doesn't mean there was anything living in the meteorites; there wasn't. It just meant that even in the absence of life, these substances form—on the way to life.

In other words, it is not just in laboratory experiments that chemical changes seem to move in the direction of life. It also happens in meteorites where there is no human interference or direction at all.

There is one more place where interesting results can be obtained. These are the vast clouds of dust and gas that are to be found between the stars in various parts of our galaxy.

These clouds of dust and gas (similar to that from which the solar system was formed) are many trillions of miles away, but they can be studied by means of the radio waves they send out. Every substance sends out radio waves, and every different kind of molecule sends out a different combination of radio waves. Each molecule has its own radio "fingerprint," so to speak.

It wasn't till the late 1960s, however, that human beings developed *radio telescopes* sufficiently advanced to collect these faint radio waves and to analyze them properly.

In 1968 the radio-wave fingerprints of water and ammonia were detected in these dust clouds. Then,

49

RADIO TELESCOPE

in 1969, the first carbon-containing compound was detected—formaldehyde.

All through the 1970s more and more compounds were detected that were more and more complicated. Almost all of them were carbon-containing, and some of them contained up to seven or eight atoms apiece.

The English astronomer Fred Hoyle (1915–) suggested that there might even be small quantities of proteins and nucleic acids formed in such clouds. These might be too small for us to detect, but they might represent life. Perhaps that is where life started, and perhaps life reached Earth from such clouds.

This is not a very likely suggestion, but scientists are only at the beginning of their attempts to find out how life began. Considering how long, long ago it must have happened and what faint clues there are, it is surprising that they have been able to work out as much as they have.

In future years they will do a great deal better still.

Index

Index